Norway Coastline Dusk Fjord Sea Mountain Snow

Ship Vacation Cruise Cruises Norway

Fjord Norway Mountains Landscape Rock View

Nordic Norway The Fjord

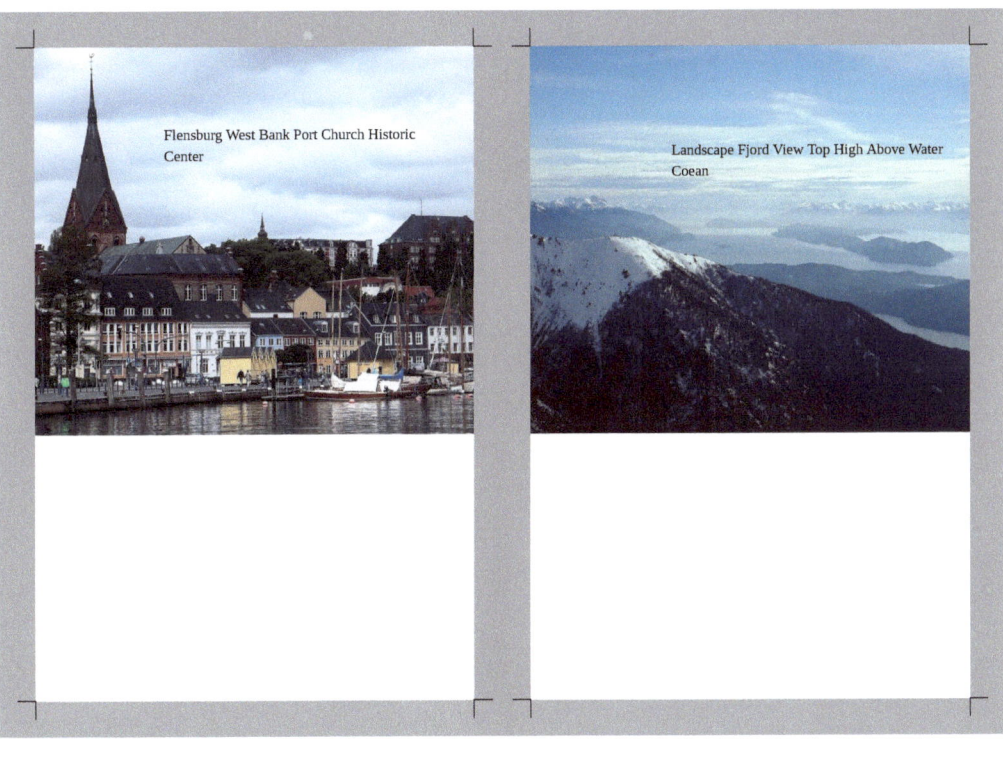

Norway Coastline Ship Fjord Sea Mountain Snow

Predigtstuhl Cliff Norway Fjord Mountain Nature

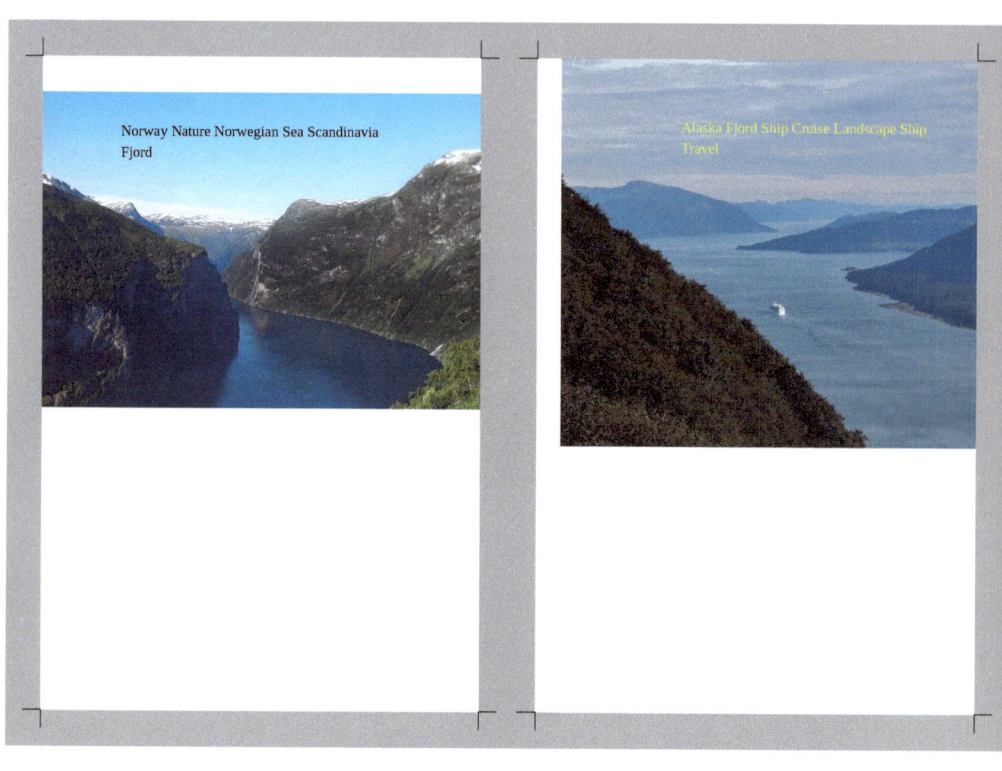

Norway Nature Norwegian Sea Scandinavia Fjord

Alaska Fjord Ship Cruise Landscape Ship Travel

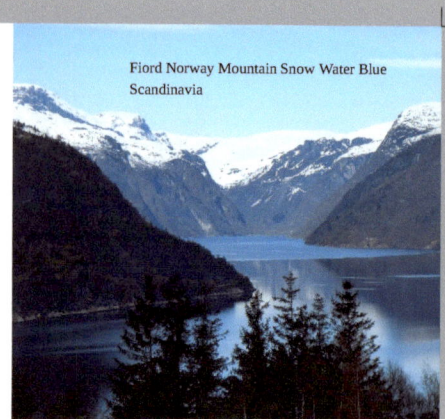

Fjords Norway Landscape

Landscape Norway Fjord Travel View Snow Panorama

Iceland Cliffs Basalt Fjord

Breen Glacier Fjord Norway Water Lake Mountains

Alesund Norway Cruise Fjords Port Holiday Cruise

Preikestolen Norway Rock View Fjord Lysefjord

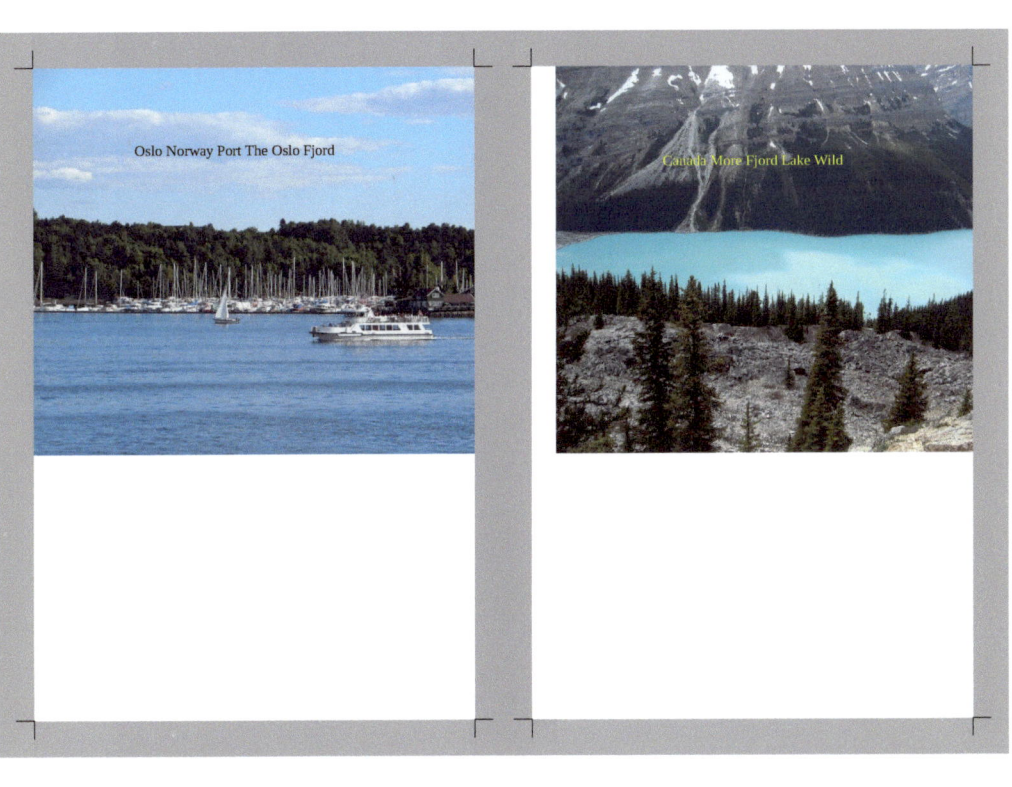

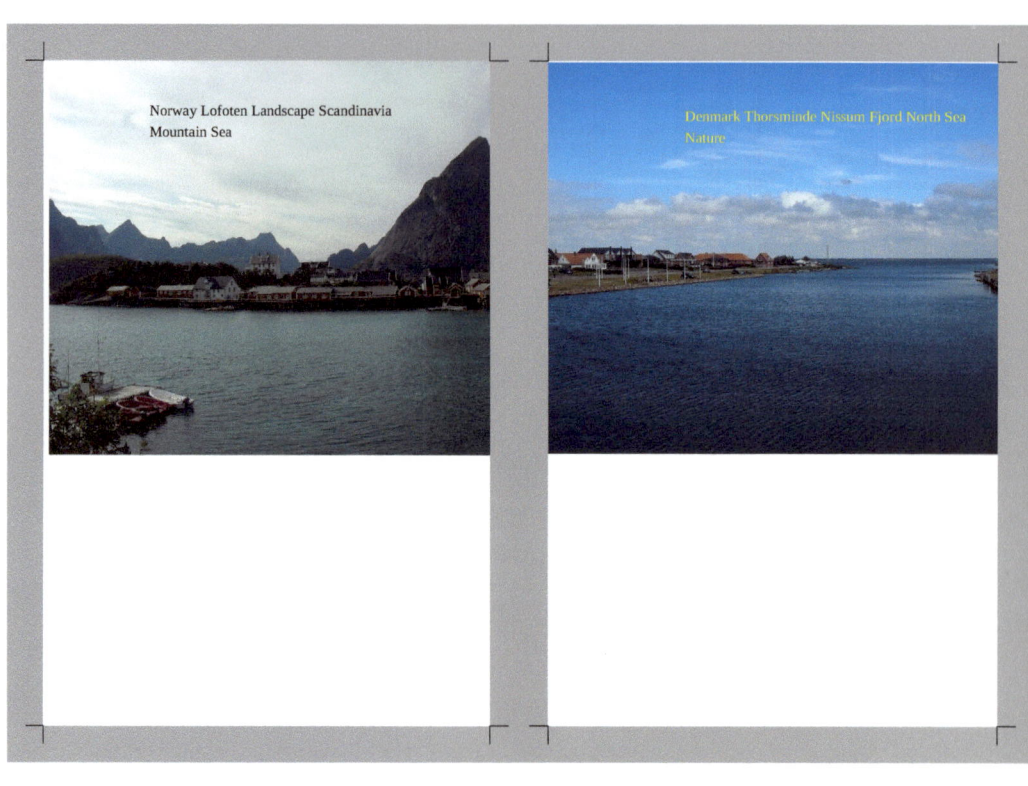

# Proof

www.ingramcontent.com/pod-product-compliance
Lightning Source LLC
Chambersburg PA
CBHW040316220526
45473CB00009B/2455